Introduction to design
with electron tubes

José Ignacio Domínguez Simón

ISBN: 9781549714184

Dedicated to Irene

Table of contents

Preface

Introduction

At present, analog electronics has reached what might easily be called a state of deep maturity. With analog systems increasingly replaced by their digital versions in engineering and consumer equipment, it seems some classic devices, such as vacuum tubes, have no place in the market. However, far from disappearing, tubes have become irreplaceable in certain fields of modern technology. Most important sectors that currently make use of them are audio equipment and power amplification for radio frequency bands.

In the whole chain needed to generate a product involving audio (music record, film, television or radio, etc.) and take it to the final destination (user equipment), we found lots of appliances that use tubes. From the singer's microphone preamplifier to the power amplifier of the stereo where the disc is being played, tubes can bring certain characteristics to the sound that -far from being discussed in this text- are usually highly sought after by professionals and audiophiles.

Nobody can deny this fact because, even in the best studios around the world where analog technology has been almost completely discarded, the only pieces of this type that remain are those that incorporate tubes. Thanks to the combination of classic and modern technology, unprecedented results are achieved.

Historically, its importance was revolutionary for being the first device that made active signal amplification possible. That, led to the expansion of the radio and the development of modern electronics.

With the popularization of the transistor, the disadvantages of the tube in size, fragility, weight, efficiency and price, caused its market share to be reduced to a marginal level.

The world of sound has always maintained a significant level of use as, at some tasks, tubes are considered irreplaceable. However, being an obsolete technology, usually, maintenance and repair of tube equipment is unavailable or ineffective, largely because of the arduousness of finding dedicated professionals (it is not being taught in the universities for decades now). Today, most engineers do not know tube systems deeply enough.

Tolerances

Manufacture of electronic tubes has fallen since the introduction of the transistor in the fifties, only to be replaced almost entirely in the sixties. This exclusion only escaped certain sectors where, despite its drawbacks, the advantages provided were essential (radio receivers and transmitters and some sound equipment). Today are still made only in Russia, China and Japan, the main brands being Sovtek (Russia) and Shuguang (China). The quality of modern tubes is often discussed among audio professionals.

Tube manufacturing, unlike transistors, requires much manual control by the operator, in order to position and weld the inner parts. These parts are also considerably large (from a few millimeters to centimeters) and sensitive to vibrations, impacts and thermal variations. Additionally, the high vacuum achieved in the tube can be lost, mainly because of the difficulty of maintaining a good sealing between the glass and the metal connector pins. These features represent a problem in maintaining tolerance of manufactured units. While the process of construction of a transistor is fully automated and is performed with microscopic accuracy, for the tubes it is almost handwork with large variability. This represents the second problem: tubes at first glance identical may have quite different electronic properties, even between units from the same manufacturer and batch. Then in cases of, for example, symmetric configurations where several tubes should work almost identically at the same time, it is necessary to perform individualized measurements before these units operate.

The typical tolerance of a new tube can be around 20%. Unlike solid state devices[1], as time goes by, the internal tube materials barely change. One would expect that the tubes are very stable over the years, however, certain factors such as loss of vacuum (mainly due to the sealing between the glass and the metal pins) and the appearance of gaseous waste (as a result of electrons impacting the anode) can dramatically alter the initial operation conditions. The existence of unwanted gases inside leads to a profound and rapid degradation of the internal electrodes, mainly by oxidation and sputtering, which is accelerated by the high temperatures reached. As a result, it is common for a tube to age over time, even more if it has been used.

Aging and high intrinsic tolerance of tubes make individual analysis prior to use almost essential for some types of circuits. Most devices that use tubes in stages where the drift of its parameters is critical, have some manual control to adjust the circuit according to the tube that is installed at that time. Thus, optimizing the system is achieved regardless of the tolerance of the tubes (to some extent).

---

1   Feinberg et al., (2005). Parametric degradation in transistors. *Reliability and Maintainability Symposium. Proceedings. Annual.*

This raises the next problem; testing the tubes requires specialized gear.

When tubes represented the heart of electronic devices, several types of measuring tools were easily available. With the decline of its hegemony, they were gradually replaced by transistor meters to the point where, today, it is almost impossible to get one. And the existing units are old, have been used for decades, and contain internal components that may offer totally out of precision measurements. Needless to say, all of them are analogue and many can only carry out simple analysis providing very broad tube characteristics. These old instruments are widely used by sellers of tubes, as well as designers and advanced users. In most cases, they only incorporate a needle indicator that outline the "report" of the tubes in a single number. In recent years, some companies have produced digital tube meters, but have not had a wide market spread. In no case have come to be manufactured on a large scale as happened with traditional meters.

Electronic tube meters

A more thorough analysis that can be performed to a tube is to sweep and trace its curves family under typical operating conditions. Since they are non-linear devices, the curves must be sampled as accurate as possible in different configurations. Meters that perform this function are called curve tracers. As yet, the only meters marketed worldwide that perform this function are Tektronix models 570 and 575. Although the latter was designed to draw transistor curves, an adapter socket (usually homemade) allows the use with tubes. The 570 model was discontinued in 1966 and the 575 in 1971. The few units that can be found on the used market have prices usually over one thousand euros. Given their age, even if well preserved, they will require replacement of many parts and a full adjustment to meet specs. It should be noted that these meters used tubes internally, which in turn represents another problem with regard to their maintenance. With this scenario in terms of meters, there is currently no manufacturer that provides modern, accurate, fairly priced meters.

Types of tubes and their operation

Introduction

Electronics, radio and telecommunications owe much to vacuum tubes. The invention of the triode by Lee De Forest in 1906 made possible the electronic amplification of signals and led to the expansion of radio communications on a large scale. With this arrangement, a weak signal could be used to obtain a higher level replica. Previously, the methods that existed for this functionality managed to mechanically reproduce electric or acoustic signals very inaccurately. The appearance of the triode can be considered the official founding of modern technology.

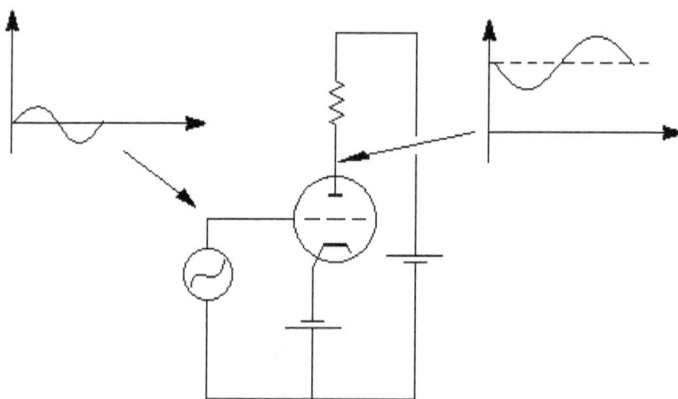

Fig. 1 – Basic idea of vacuum tube amplification. Triode tube represented with three electrodes inside a circumference; anode (top), control grid (central, discontinuous) and cathode (bottom). From a low level signal a larger one is obtained. The triode was the first active electronic device that allowed the amplification of electrical signals. Note the bias applied.

A tube is a device through which a current flow can be modulated by a control voltage. The latter signal is actually amplified. If the tube is connected in series with a load and a power source, the voltage drop in the load will depend on the current in the circuit. Thus if the control voltage is varied, the drop in the load will do according. If the voltage drop across the load is greater than the control voltage, the electronic amplification is accomplished. This is surely the most important use of the triode.

Diode and triode tubes:

The operation of tubes is based on the thermionic effect. Metals have some electrons able to move between different atoms, occasionally even escaping the material spontaneously. When the metal is heated, the speed of electrons is increased, enabling many more to escape. This phenomenon is used in the tubes, where an alloy of metals prone to lose electrons is heated (typically by a red hot filament) and placed close to a cold electrode that can receive them. A voltage is applied between these electrodes, creating an electric field that directs the free electrons.

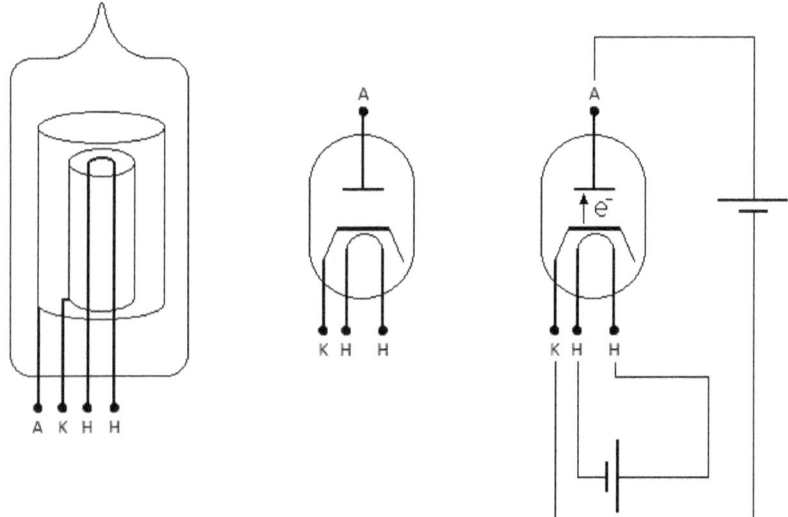

Fig. 2 – From left to right: structure of an indirectly heated diode, symbol used in a circuit, electrons flow occurs when the anode is made positive with respect to cathode. Note the cathode heating circuit.

It thus allows current to flow in a single direction (the hot electrode, cathode, can lose electrons to the cold one, anode). The device described is called diode (two electrodes between which a current can flow in one direction). If it includes a third electrode (control), then it is a triode. The potential of the third electrode modifies the existing electric field and thus alters the current (amount of electrons traveling per time unit).

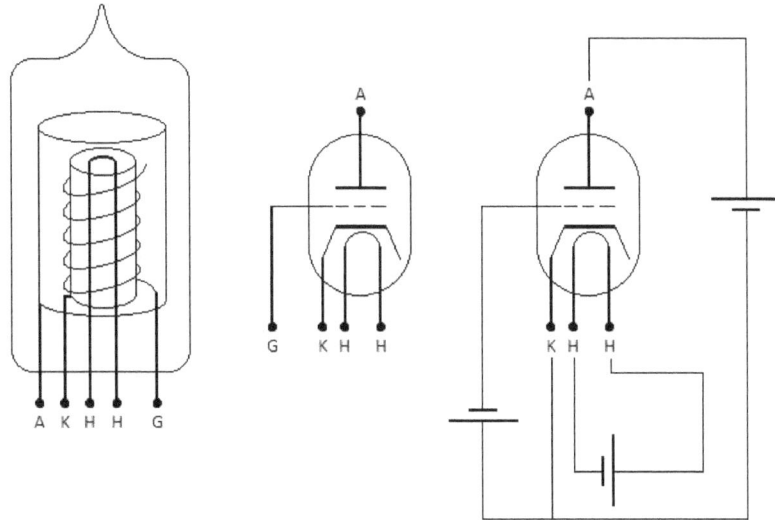

Fig. 3 – From left to right: structure of an indirectly heated triode, symbol used in a circuit, electrons flow occurs when the anode is made positive with respect to cathode. The current that is developed between the anode (A) and cathode (K) is dependent on the voltage between grid (G) and cathode.

Because the control electrode (grid control) does not form a closed loop, if kept at a lower potential than the electron emitter (cathode), the current flowing through this is, in first approximation, zero. This property gives the tube a very strong ability to amplify signals from sources capable of delivering very low current, because the input signal (to be amplified) attacks a circuit with nearly infinite impedance. This operation mode is called negative grid bias since the control grid is maintained at a negative potential relative to the cathode, and is commonly used in amplifier circuits.

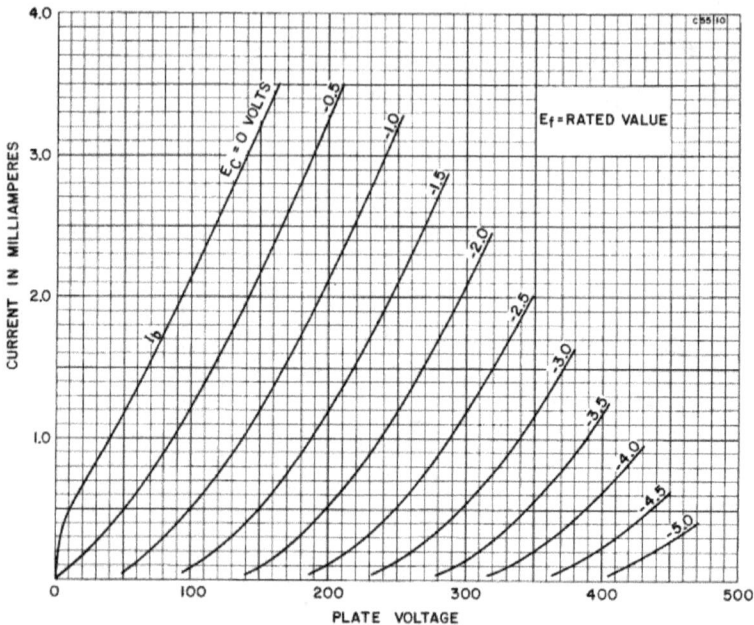

Fig. 4 – Typical curve set of a triode. Source: data sheet of the dual triode 12AX7 from Sylvania.

Tubes are usually constructed as a glass capsule with the electrodes disposed therein on a structure (typically mica) that keeps them motionless and avoids coming into contact. The capsule is sealed, under vacuum in most cases, to prevent electrons to collide with other particles in their path -this would cause them to deviate from its path and reduce the performance-. Besides, physical damage would occur; since the shock can generate positively charged ions that would bombard the cathode. Therefore, it is most common to remove the air and permanently seal the unit. There exist some special tubes filled with a gas that, when ionized, confer different performance characteristics.

The tetrode tube:

Two conductors separated by an insulating material constitute in any case a capacitor. Thus, the electrodes of a tube form capacitors whose insulator is (typically) vacuum, and with values that are called "inter-electrode capacities". The magnitude of these reactances are measured with the tube being disconnected and cold, as with any common capacitor. Under certain working conditions, the effective value of some of the internal capacities can be many times higher than measured. This phenomenon was discovered by John Milton Miller in 1920 and is responsible for the most important electronic limitation of the tubes; the Miller effect. Because reactance between electrodes are formed, unavoidable current flowing occur. In typical circuit topologies, it results in a feedback path that effectively limits the maximum working frequency. That is, reducing the actual electrical separation between the electrodes.

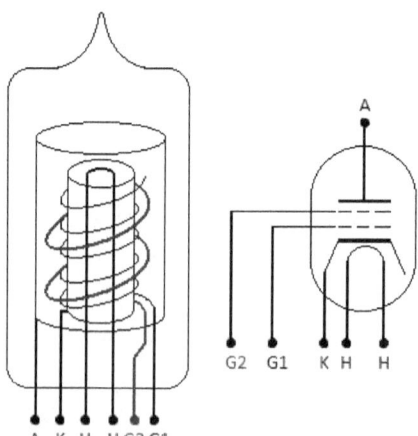

Fig. 5 – From left to right: construction of a indirect-heating tetrode, symbol used in a circuit. Screen-grid is named as grid number two (G2) and control-grid as grid one (G1).

To overcome the limitation that the Miller effect introduces the tetrode was created; a tube that includes an additional electrode (screen grid) serving as electrostatic screen between the control-grid (control electrode) and the anode. This new electrode is connected to a positive potential, thereby reducing the grid-anode capacity value; usually between ten and one hundred times less than for a triode. This device is called a tetrode for having four electrodes inside. By drastically reducing the control-grid–anode capacity, the coupling between the two electrodes is much smaller and the separation between their respective circuits improves, ultimately allowing the tube to operate at higher frequencies.

The secondary emission:

The introduction of this grid adds a drawback; the acceleration to the electrons produced by the additional (positive) grid leads to – when the anode current (cathode-anode) becomes large enough– the electrons reaching the anode with sufficient energy to tear other electrons from it. The latter would go back to the anode in a triode or diode, but in the tetrode they may end up in the screen-grid (if it has a higher potential than the anode), reducing thereby the anode current and introducing another important non-linear characteristic of this tube.

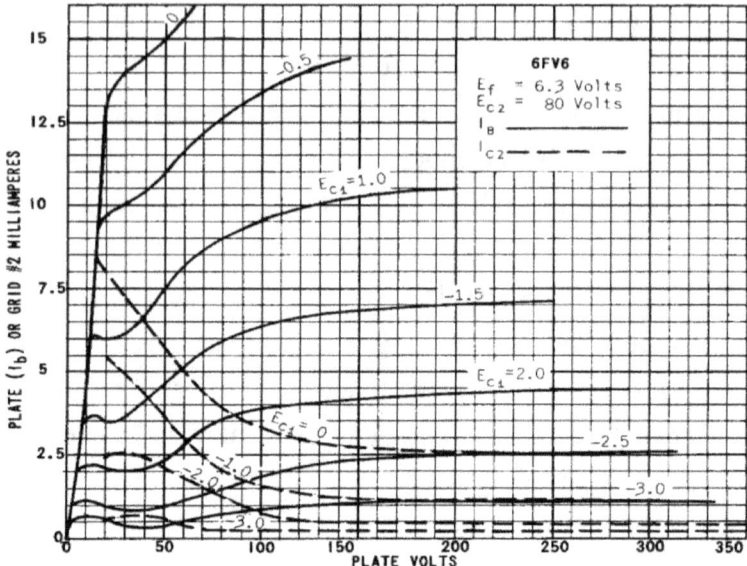

Fig. 6 – Typical family of curves of a tetrode. Notice the 'kink' that appears in the curves for low values of anodic voltage. The cause is the secondary emission caused by the impact of high kinetic energy electrons at the anode, which end up in the screen grid, causing a decrease of the anode current. When the anode reaches a sufficiently high voltage, the electrons from the secondary emission are attracted back to the anode and the *undesirable* effect disappears. Importantly, the place of the curve where the elbow appears depends on the screen grid voltage display. In this figure it is connected to a DC source of 80 volts. The effect of "negative resistance" (for a small portion of the curves, the higher the anode voltage the lower the anode current) disappears when the anode voltage is higher than the screen grid. Source: datasheet of the tetrode 6FV6 from the Tung-Sol brand.

This defect makes the tetrode unusable for applications where the working area is such that the anode voltage drops enough to enter the 'kink' of the secondary emission.

The beam tetrode tube ("kinkless tetrode"):

The problem with secondary emission intrinsic to this design finds a solution with the invention of the directed beam tetrode or simply 'beam tetrode'. It incorporates deflector plates, located between the anode and the screen grid and connected to the cathode. These new plates are incapable of emitting electrons (they are made of a different material and colder) but are suitable to ensure that those leaving the anode (secondary emission) are repelled back to it, preventing them from reaching the screen grid (and thus producing undesired leakage current). The 'kink' produced by secondary emission is eliminated from the characteristic curves.

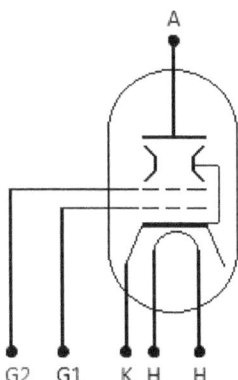

Fig. 7 – Symbol of an indirect-heating directed beam tetrode. Deflector plates are often connected to the cathode internally. Some models have them accessible through an additional socket pin to allow the designer to make use of plates independently as with the other grids.

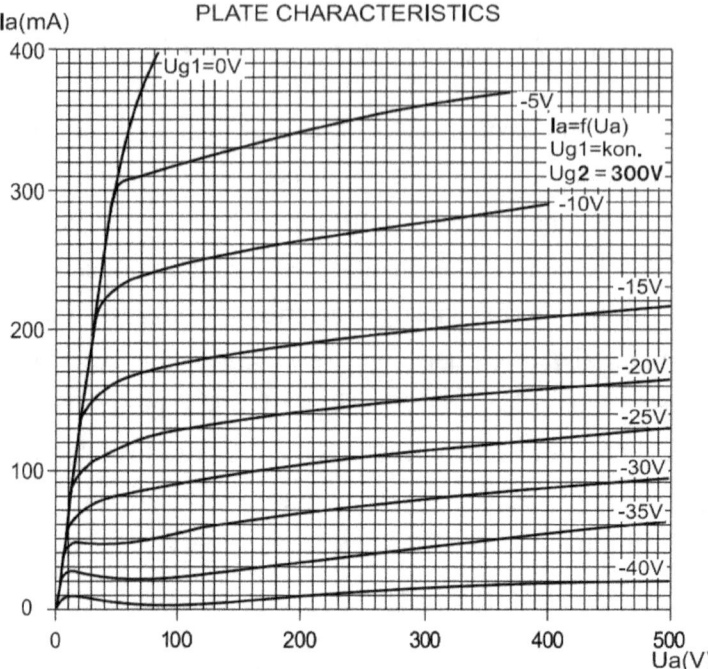

Fig. 8 – Family of typical curves of a directed beam tetrode. The effect of secondary emission is substantially reduced by the use of deflector plates. The result is very similar to that obtained by a pentode. Source: datasheet of directed beam tetrode KT88 from JJ brand, generally regarded as one of the highest quality choices for the final stages of an audio amplifier.

Tetrodes have some characteristics very similar to those of the pentodes, with a large electrostatic insulation between anode and control grid (low grid-anode capacity) and high gain over a wide range of anode voltages (due to the large acceleration of electrons produced by the -usually constant- high voltage of the screen grid).

The pentode tube:

The problems introduced in the tetrode by incorporating an additional grid (screen grid) can also be solved by adding another one; the suppressor grid. It is physically arranged between the screen grid and the anode and is connected to a lower potential than the anode (typically to the cathode), preventing electrons detached from the anode from ending up in the screen grid. It effectively suppresses the secondary emission (emission of electrons from the anode that would flow to the screen grid), therefore called the suppressor grid. Unlike the cathode, it is not made of a material with great ability to lose electrons when heated, so it acts as a control grid and not as an electron source.

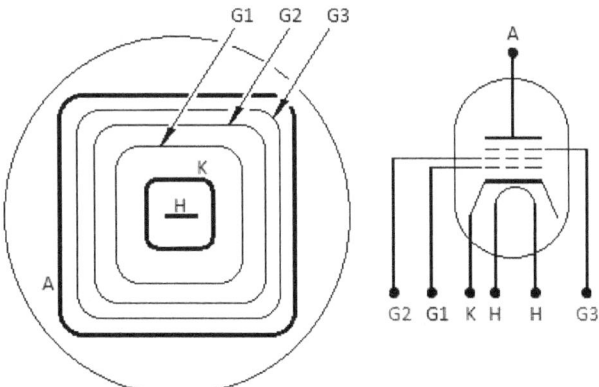

Fig. 9 – From left to right; vertical sectional view of a pentode (with the order of electrodes) and symbol used in a circuit. Nomenclature of the electrodes: anode (A), cathode (K), control grid (G1), screen grid (G2), suppressor grid (G3).

With all this, thanks to the screen grid, the effect of inter-electrode capacities (most importantly that formed between the control grid and the anode) and the anode current leakage are minimized. The pentode tube is the most efficient and improved type in terms of signal amplification. This is the main reason why it is usually preferred for the output stage of power amplifiers and also for amplifying weak RF signals.

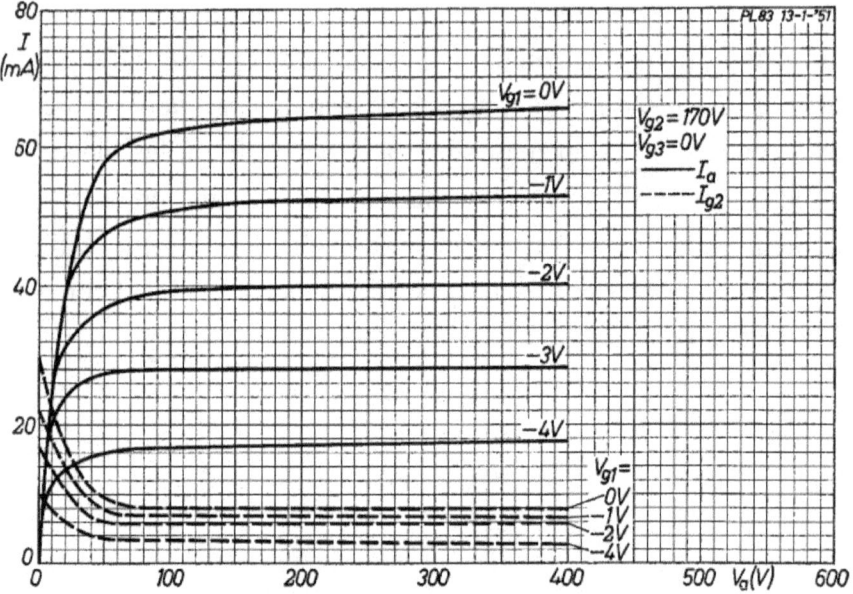

Fig. 10 – Typical family of curves of a pentode. The characteristic kink of the tetrode is removed and for values of medium to high voltage the behavior is almost that of an ideal current source (regardless of the anode voltage, the current only varies depending on the grid voltage). The reason for this comes from the fact that the electrons acceleration is mainly dependent on the screen grid voltage, which is constant. Source: datasheet of EL83 pentode, Phillips brand.

Conclusion note:

Despite the above, pentode tubes are not superior to the triodes and tetrodes in all their features. An example is the distribution of harmonics introduced (harmonic distortion products that appear in an amplification circuit when using these tubes) which is such that more energy is distributed among third-order harmonics, while for tetrodes there is more energy found in second order harmonics. Thus, in a push-pull configuration, tetrodes will produce an output with less distortion (being the second order harmonics largely canceled due to the circuit topology). When it comes to radio, this distortion in the output may be negligible depending on the case, but if applied to the audio sector it is a critical and sometimes excluding factor.

Triodes, despite their clear disadvantage to tetrodes and pentodes for their inter-electrode capacity, are the most linear tubes within these family. That is the reason why they are preferred in many electronic devices where high fidelity is required for signal processing or amplification.

Tubes characterization

Typical family of curves of a triode

The behavior of a triode is, relatively speaking, somewhat similar to a resistor, although much less linear. Plotting the V-I (voltage vs. current) curve corresponding to an ideal resistor of, for example 20kΩ, shows the following:

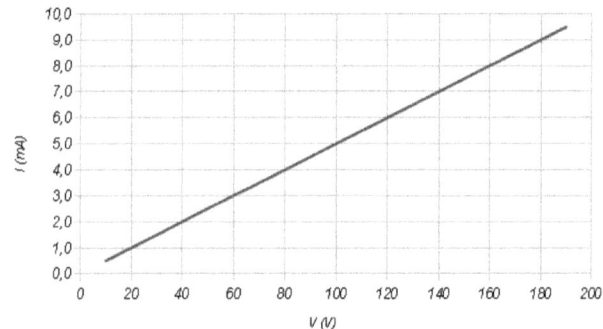

Fig. 11 – Characteristic curve voltage – current of an ideal resistor. The linear behavior clearly allows the use of a simple equivalent model. La resistor is a passive element, then the curve inevitably pass through the origin.

The slope of the line, defined as $\dfrac{\Delta Y(A)}{\Delta X(V)}=C(S)$ [2] corresponds to the value of the conductance in Siemens (inverse of resistance) of the element. In the case of an ideal resistor, being a linear element, measuring the slope at any point on the line, it is possible to know all others. The calculation thereof would be $y=C\cdot x=R^{-1}\cdot x$ [3],

---

2 The characteristic curve has conductance unit. It applies to any device under test from which current-voltage curve is being traced.

3 Ohm's law expresses relation between magnitudes in both axis of a characteristic curve plot, relating voltage to current (impedance).

corresponding to Ohm's law. It is due to the device being linear (in low frequency) that such a simple characterization can be used.

When evaluating the behavior of a triode, its nonlinearity character prevents from using such a simple model, since the curves are not described by simple equations. One possible solution is to measure the curves at many points to get an accurate representation of the actual behavior.

Drawing from the data sheet specifications of a common model triode (12AX7), a typical family of curves is similar to that shown in Figure 12.

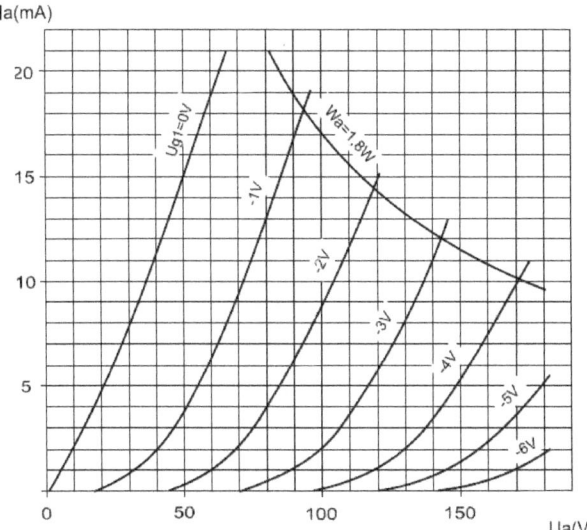

Fig. 12 – Family of characteristic curves of a typical triode. Source: datasheet of triode model 12AX7, brand JJ.

Observing the family of curves, three particularly important details are perceived:

- the curves correspond to a negative voltage (grid – cathode) -that is the parameter that must be changed to obtain different curves-,

- for low values of anodic current (vertical axis), the behavior is more non-linear,

- linearity varies across different curves, being more linear those corresponding to voltages of grid-cathode closer to zero,

- curves are not parallel and the slope of some (those on the left) is more steep than others (on the right)

At first glance it appears to be a too non-linear device, but if the nominal area of operation is restricted to a centered region on the curve segment, the mistake for considering it a linear device can be small or even negligible. Therefore, choosing the proper operating point (bias) is a decisive factor to achieve good results.

Process of sweeping the curves

Finding the family of curves of a triode is an operation that can be performed without specific gear. Just with the help of a power supply capable of delivering voltages in the operating range (typically from 0 to 300 volts or more) and using a milliammeter to measure the anode currents developed in each case. By applying different voltages, and measuring their corresponding currents, curves can be drawn from the value pairs obtained. This method, although valid, is not desirable; it would render the whole process very slow.

Modern characterization systems use short duration pulses that vary the polarization throughout the useful range and measure the current-voltage pairs for each curve of the tube. An example of this is the solution developed by the author; the TMS, a computer controlled system, capable of accurately characterizing a tube in seconds with less than 1% error.

Introduction to the equivalent model

If the range of interest is reduced to a central area where the curves have a desirable behavior, the tube can be considered a linear system. It is important to remember that the model assumes small variations compared to the complete length of the curve. Beyond that, the obtained behavior will not meet the calculations, simply because the assumptions made correspond to a linear system.

For example, in the case in which, by design, the work area is that indicated in Figure 14,

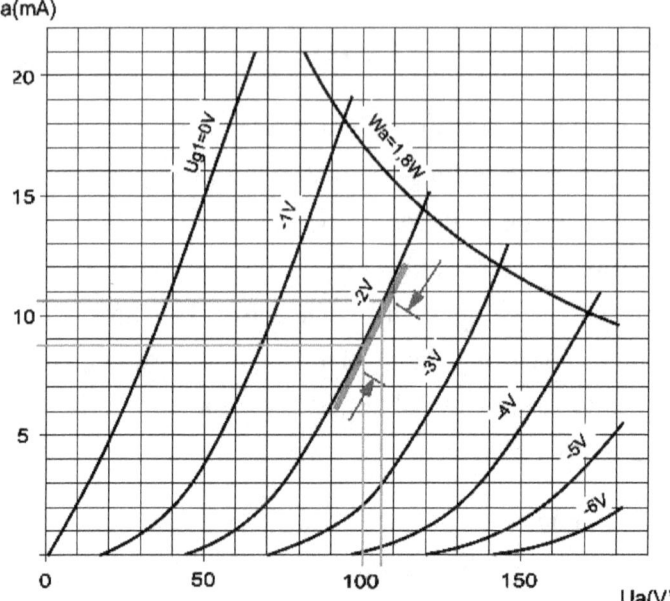

Fig. 14 – Estimation of the equivalent internal resistance in a triode from the slope of the curve. If the operating range is reduced to a small part, the curve can be considered a straight segment.

whose behavior can be equated to that of an ideal resistor, with a value of:

$$\frac{107V - 100V}{10,8\,mA - 8,9\,mA} = 3684\,(\Omega) \quad [4]$$

which would be obtained with a grid bias of -2 Volts (since the curve corresponds to a grid – cathode voltage of -2 Volts). AC voltages are usually also applied, which will result in a variating output centered at the bias point:

---

4 Curve tangent must be calculated from a reduced segment to avoid introducing excessive error.

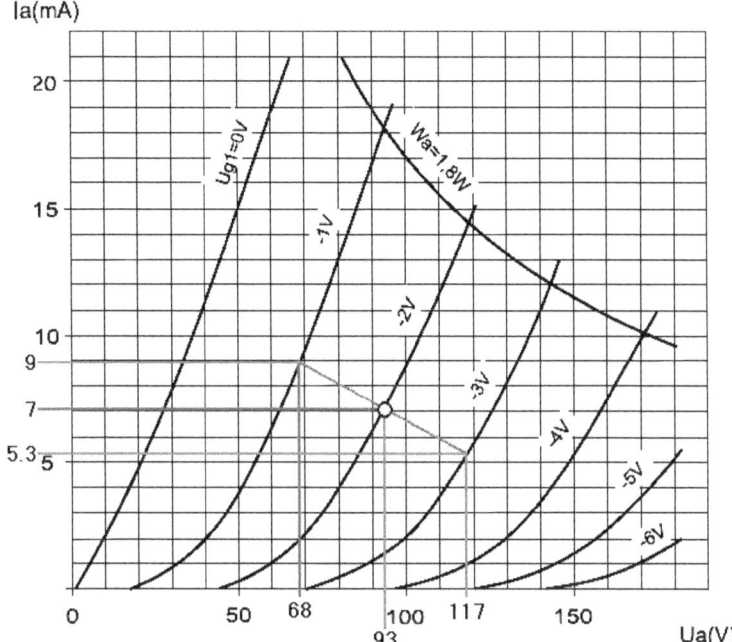

Fig. 15 – The voltage gain of the circuit can be obtained from the intersection with adjacent curves. For that, knowing the 'load line' is needed, which is function of the real part of the anode load impedance and anode voltage supply.

that, using the proper circuit where output voltage relates to input voltage (grid – cathode voltage), translates into a gain following equation 1.

$$G = \frac{117\,V - 68\,V}{(-3V) - (-1V)} = -24,5\,(V/V)\ ^5, \qquad [1]$$

In this case the gain would be negative (the output is inverted with respect to the input). In a typical circuit, the working conditions are carefully chosen to restrict the tube operation to some segments of some curves, making the equivalent model accurate enough (figure 16).

---

5   Voltage gain estimated with chosen polarization.

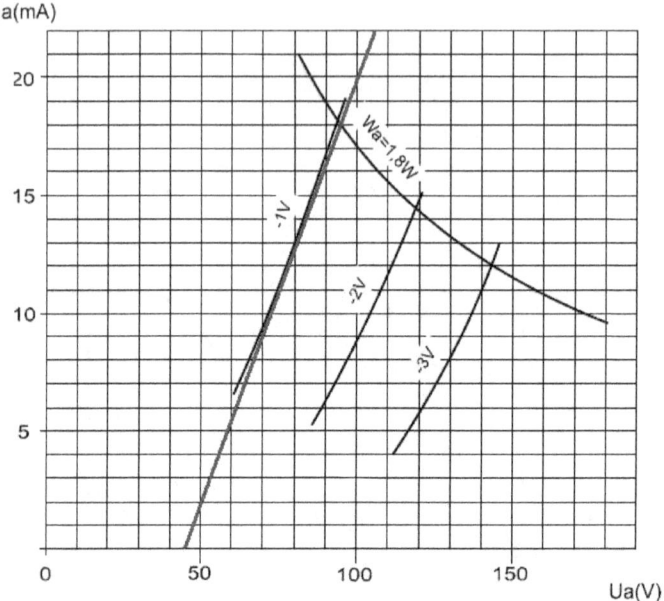

Fig. 16 – Similarity between equivalent model and real curve of a triode. They can be considered identical for a central segment of the curve.

The tangent straight line has being generated from the equivalent model. The difference with real curve (provided by the manufacturer) is acceptable as long as the work region does not greatly exceed the previously considered limits.

The most commonly used equivalent model consists of a voltage controlled resistor (function of anode voltage) in parallel with a current generator controlled by the grid-cathode voltage:

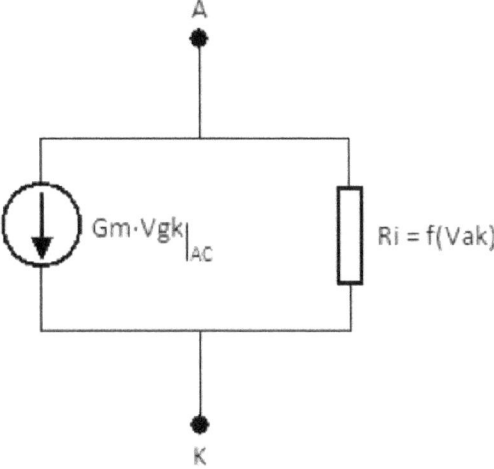

Fig. 17 – Equivalent model, formed by a voltage controlled current generator and a resistor whose resistance depends on the bias point (anode voltage and current).

Explanation of parts:

The resistor models the behavior of the tube at that bias point (central point in the curve) and it is the voltage – current relationship that can be calculated from the slope on the plot.

The voltage controlled current source models the variation of the current through the tube when the grid-cathode voltage is varied around the central (bias) point.

The transconductance of a triode is defined as the variation of current that the anode experiences when grid-cathode voltage varies while keeping a constant anode voltage.

One possible test circuit to measure the transconductance following such definition could be:

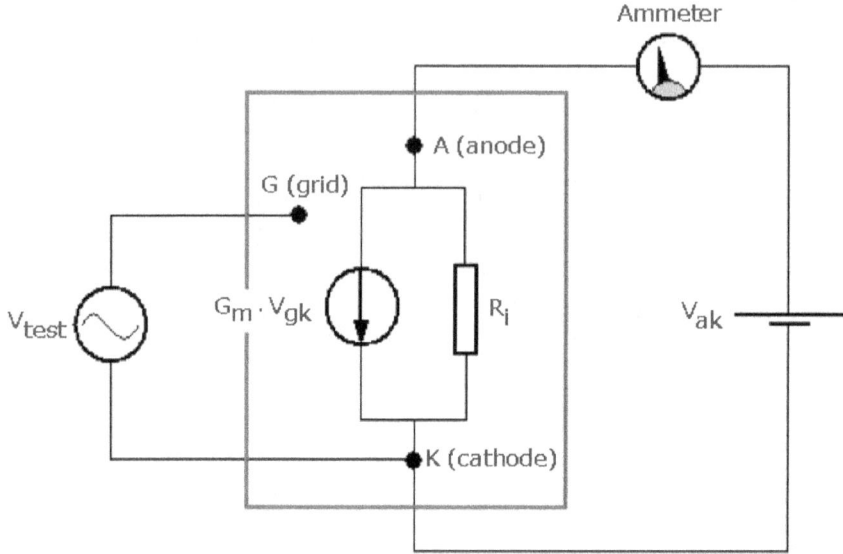

Fig. 18 – Basic idea for characterizing a triode under test. Curves sweeping is conducted by varying the voltage applied to the grid-cathode electrodes and measuring the developed anode current. Being a non-linear device controlled by an external voltage, it is necessary to perform multiple sweepings (therefore obtaining parametric results).

Keeping anode-cathode voltage constant by means of V_{ak}, the current variation is measured for different values of the grid-cathode voltage. Conducting several tests while sweeping through V_{ak} values, the parameters of the equivalent model can be measured for all different bias setups.

Note: in first approximation, the grid is an open circuit, thus no current flows. Indeed, for low and medium frequencies in negative bias, the very weak currents that flow are negligible. If

working with positive bias, the current will become much larger (from microamps to milliamps), because the grid is at positive potential relative to the cathode and attracts the electrons it emits. Normally this is not a desirable working condition for low-power tubes (which can also lead to the destruction of the grid due to overheating).

Linear equivalent model:

A tube is an active electronic device capable of controlling a current according to an input voltage. Far from being a linear element in low-frequency as typical passive components (resistors, capacitors, inductors), the tubes are distinguished for being very little linear over all range of useful frequencies. However, if its application in a circuit restricts the operation area to a relatively narrow range of voltages, tubes can be considered a quasi linear device. The exact behavior of the tubes is most accurately defined by the family of curves; a graph where the current-voltage relationship between anode and cathode of the tube for different polarization cases is shown. Conducting a graphical analysis is the most accurate way of estimating the behavior that a tube will have on a circuit. This method introduces no other error besides those that might appear while getting the coordinates of a given point, that is, the graph precision itself. Performing a graphical analysis easily provides the static conditions (bias) of the tube circuit with good accuracy. Also from the curves, the parameters of the linear equivalent model can be obtained, although those are valid only as an approximation, within a certain range from bias point. The linear equivalent model helps defining the tube behavior when AC signals are present.

Polarization and internal parameters:

Basically, the polarization of a tube consists of applying a current-limited constant high voltage supply (typically 100 to 500 Volts) between anode and cathode, while maintaining the grid voltage below that of the cathode. This allows a certain DC current to flow through the anode – cathode circuit, with a value given by the family of curves for the corresponding bias point. This current is commonly referred to as "bias current", and is the value that can be measured when no alternating signals are present; e.g. with no input signal in the case of an amplifier. The voltage at which the grid is clamped is the parameter that decides which curve of the family is "active".

The current – voltage ratio between anode and cathode responds to a "virtual resistance", usually "anode resistance" or "internal resistance" (found as R_i or r_i), whose value (Ohms) depends on the bias point (slope of the curve in each case). It is therefore modeled as a voltage / current controlled resistor, a property usually undesired but that in some cases can be used to design a circuit that needs this non-linear behavior.

Grid voltage will subsequently vary according to the AC signal to be amplified while maintaining the DC component which was set with the chosen polarization. This instantaneous variation (AC), typically of small amplitude relative to the polarization (DC), will cause a fluctuation of the anode current accordingly. The ratio with which these two quantities vary is called transconductance or mutual conductance (G_m), and its unit is the amp / volt, i.e. in how many Amperes the anode current grows per each Volt increase of the grid voltage. Often, the submultiple milliamp per Volt (mA / V) is used instead, since it suits better is the typical working range.

The values of internal resistance and transconductance vary depending on the specific curve and the curve's region, so a single value defining the tube can not be given. However, both vary according to a defined relationship; if the parameters of internal resistance and transconductance multiplied, the amplification factor (μ) of the tube is obtained. This reflects the ratio of the variation of the anode voltage with the grid voltage (Volt / Volt), and remains fairly constant in virtually all the working range of the tube. This is the reason why it is often provided in datasheets as a defining characteristic regardless of the working conditions.

$$\mu\left(V/V\right)=G_m\left(mA/V\right)\cdot r_i\left(\Omega\right) \quad [6]$$ [2]

Approximating the tube as a linear circuit, the equivalent model comprises two components;

- the internal resistor, of variable value depending on the bias setting (DC) and

- a voltage-controlled voltage source, depending on the grid-cathode voltage (AC).

These elements would be in series, forming a real voltage generator, similar to a Thevenin equivalent circuit. They can, likewise, be interpreted as a real current generator, in that case counting:

- internal resistance, same value as before, and

- a voltage-controlled current generator, dependent on the grid-cathode voltage.

---

6 Relationship between quantities characterizing a triode; amplification factor (μ), transconductance (Gm) and internal resistance (ri).

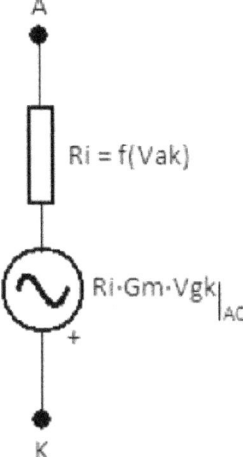

Fig. 19 – Schematic of the equivalent model with a voltage-controlled voltage generator. The internal resistance depends on the polarization (grid-cathode and anode-cathode voltages). The generator responds to changes in the AC grid voltage. Therefore, if no variable-excitation ($V_{gk|AC} = 0$) is applied to the grid, the tube behaves as a resistor whose resistance depends on the bias point (diode).

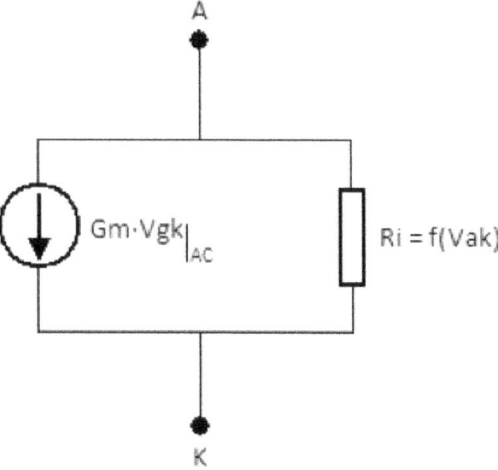

Fig. 20 – Schematic of the equivalent model as a voltage-controlled current generator. When no alternating excitation of the grid component is present, the current can only flow through the internal resistance (with a value that depends on the polarization).

Equivalent model application:

The use of either model is irrelevant, since they are equivalent circuits. The proposed model is always valid, but it is important to remember that the values of transconductance and internal resistance are not constant, so for an accurate calculation, they should be considered as variables rather than constants in the equations of the circuit analysis (a way to combine the analysis of the equivalent linear model with variations arising from the characteristic curves).

Typically, the amplitude of the AC signal applied to the grid is small compared to the bias value, so if the working point is located in a very linear region of the curve, the parameters can be considered constant, thereby introducing a minimum error.

Getting the parameters:

From the family of curves of a tube, the equivalent model parameters can be obtained based on:

- the slope of the curve (internal resistance r_i),

- horizontal separation between curves (amplification factor, μ) and

- the vertical separation between the curves (transconductance, G_m).

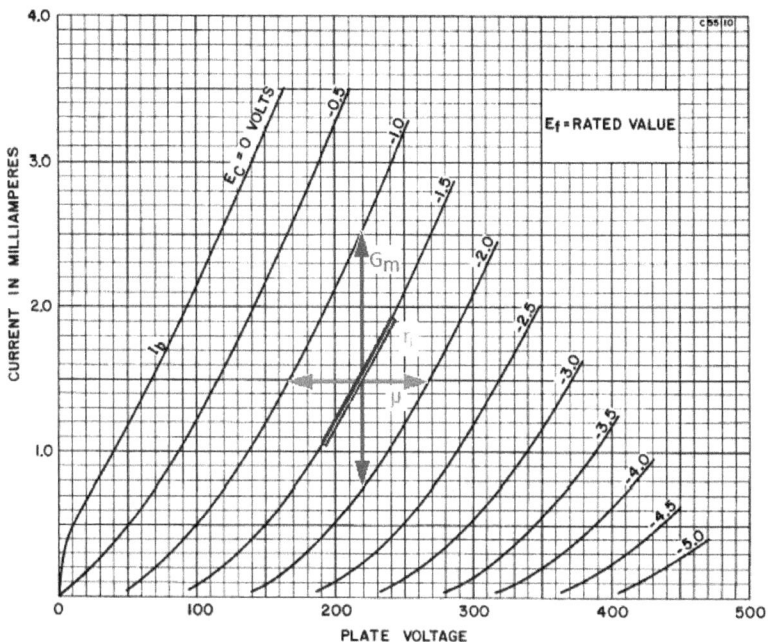

Fig. 21 – Obtaining internal parameters of equivalent model from family of curves.

Following the former definitions, the internal parameters are raised with the following assumptions:

−   Internal resistance: slope of the curve in a limited area.

−   Amplification factor: ratio of variation of anode voltage to grid voltage maintaining anode-cathode current constant.

−   Transconductance: variation of the anodic current respect to the grid voltage keeping the anode-cathode voltage constant.

These calculations can be performed manually on the family of curves or automatically if the points of the curves are available in digital format (pairs of coordinates current-voltage anode-cathode).

Inter-electrode capacities and Miller effect:

Origin:

The construction of a tube means usually having a number of very close electrodes but without physical contact, separated by vacuum. Between these an electronic current can flow. This current originates from a effect known as thermionic.

Amplification is possible when the flow of electrons can be modulated externally, by using a "control" electrode, that physically stands in the existing flow, having the ability to reduce or increase it by changing the electric field present. The performance of the control electrode depends to a large extent on the proximity to the electron-receiving element (anode), to be able to deflect the traveling electrons. The closeness between electrodes is therefore inevitable to ensure the effectiveness of the system.

Following the formal definition; a capacitor is formed by two conductive elements next to each other and separated by a dielectric. A (thermionic) tube electrodes consist of metal parts separated by vacuum. Thus intrinsically, it will always form capacitors between its electrodes.

The formation of these internal capacitors would not present a problem if it was not the case that (usually) different electrodes belong to circuits that must remain uncoupled. As there exist capacitances between those parts, full separation of the circuits is impossible. This leads to a practical limitation in the use of tubes, since the coupling between electrodes is frequency dependent.

The problem of internal capacities was a major obstacle to overcome in the early days of modern electronics (limiting the operation at high frequencies), which found solution incorporating additional electrodes to provide adequate electrostatic isolation.

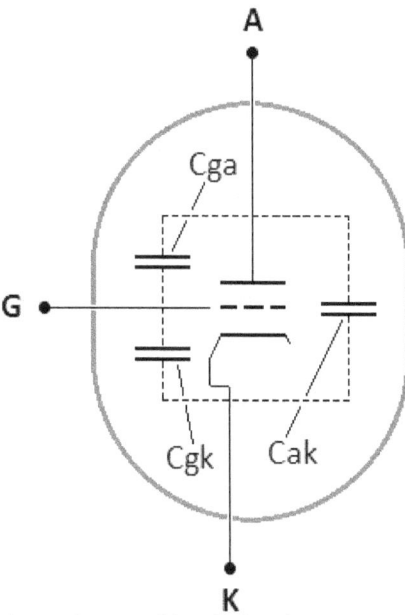

Fig. 22 – Inter-electrode capacities. Subscripts are named as the electrodes between which they are formed; grid - anode (C_{ga}), grid - cathode (C_{gk}) and anode - cathode (C_{ak}).

The values of the internal capacities, usually provided by the manufacturer, are measured with the tube cold and disconnected from the circuit, i.e., following the usual procedure to measure any normal capacitor.

Undesired effects they introduce:

The first consideration that arises is that the circuit of the control grid is not already open, and thus current can flow to the anode and cathode. This reduces the input impedance of the tube. It is usual that the input signal is applied (decoupled) between the grid and the cathode, therefore, the impedance seen by the signal source is the corresponding to the reactance of C_{gk} in parallel with C_{ga} (as the anode will be short-circuited to ground in AC analysis). But since the anode is usually in series with a high value resistor, the effect of the C_{ga} is less important and C_{gk} can be considered the main limitation. As the input signal reaches higher frequencies, the electrical separation (impedance) between the grid and the cathode decreases, reducing the effective input impedance.

For the audio frequency band, a medium-frequency signal can be considered that of around 1 kHz. For a source that produces this signal, the magnitude of the impedance seen would be, taking a typical triode model 6SN7:

$$|Z_{in}(@\,1\text{kHz})| = \frac{1}{2 \cdot \pi \cdot 1\text{khz} \cdot C_{gk}}\ {}^{[7]}, \qquad [3]$$

where capacity grid - cathode, according to the data sheet is 3pF;

$$|Z_{in}(@\,1\text{kHz})| = \frac{1}{2 \cdot \pi \cdot 1\text{khz} \cdot 3\text{pF}} = 53\text{M}\Omega\ {}^{[8]}, \qquad [4]$$

a very high value that could be useful for receiving weak signals (low amplitude) or those coming from sources with high output impedance. In the case of a source generating a 20kHz signal,

---

7   Capacitive input reactance.
8   It can be considered an open circuit for medium frequencies.

impedance would be according to Equation 5.

$$\left|Z_{in}(@\,1\text{kHz})\right|=\frac{1}{2\cdot\pi\cdot20\text{khz}\cdot3\text{pF}}=2.6\text{M}\,\Omega\ \ ^{9}, \qquad [5]$$

A considerably high value still is obtained. It is appropriate to assume that this capacity does not interfere greatly in the ability to amplify high frequency audio signals. The same applies to the capacitor formed between anode and cathode (which for high frequency signals increases the output impedance of the equivalent circuit).

The same is not true for the grid-anode capacitor, primarily responsible for the frequency limitation that circuits using tubes suffer. The explanation is that the ends of the capacitor are subjected to large potential difference, coinciding with the variation of the input signal. Thus the current that can flow through the capacitor is higher; as many times as amplification occurs in the tube[10]. This phenomenon is known as 'Miller effect'.

This capacity (C_{ga}) creates a negative feedback loop, allowing part of the amplified signal (in opposite phase) to be injected to the input. The amount of signal fed back is directly proportional to the frequency, according to the expression of the grid-anode reactance (equation 6).

$$X_{Cga}(ideal)=\frac{1}{2\cdot\pi\cdot f\cdot C_{ga}}\ \ ^{11}. \qquad [6]$$

---

9   The working frequency should be augmented considerably to obtain reduced values.

10 In other words, it would be the equivalent of assuming that the capacitor is placed between the grid and ground, while scaling its capacity according to the amplification produced.

11 If amplification does not occur in the tube, the grid - anode reactance has the

Since the signal on the anode side of C_{ga} is amplified "A" times with respect to the input, the current through C_{ga} is augmented in equal proportion. Thus, the "effective" capacitance formed between grid and anode, as seen by the grid, increases as many times as amplification occurs in the tube, according to Miller. The effective reactance is therefore:

$$X_{Cga}(effective)=\frac{1}{2\cdot\pi\cdot f\cdot C_{ga}\cdot(1+A)}{}^{12}. \qquad [7]$$

Considering that the gain typically provided by an amplifier stage implemented with a triode ranges from 20 dB to 40 dB (10 and 100 times respectively), the negative feedback loop formed poses an impassable barrier in amplifying high frequency signals.

---

normal value specified in datasheet.

12 The amplification that occurs increases the current flowing through the capacitor and therefore the effective capacity.

Design implications:

These internal capacities present a problem that can be analyzed from the point of view of the design. Taking an example amplifier implemented with a triode operating in class A as in Figure 23,

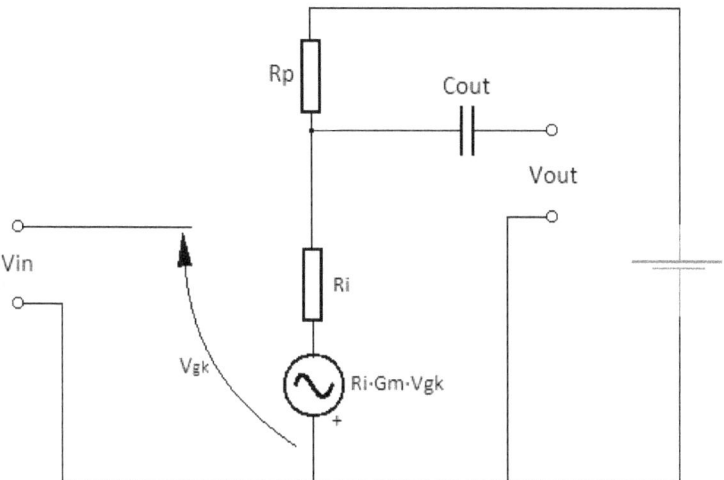

Fig. 23 – Equivalent model of a triode amplifier in class A without inter-electrode capacities.

applying the equivalent model without capacities, the circuit would be a non ideal amplifier, with output resistance not zero but infinite input impedance.

By including the inter-electrode capacities, significant changes are introduced, as shown in Figure 24.

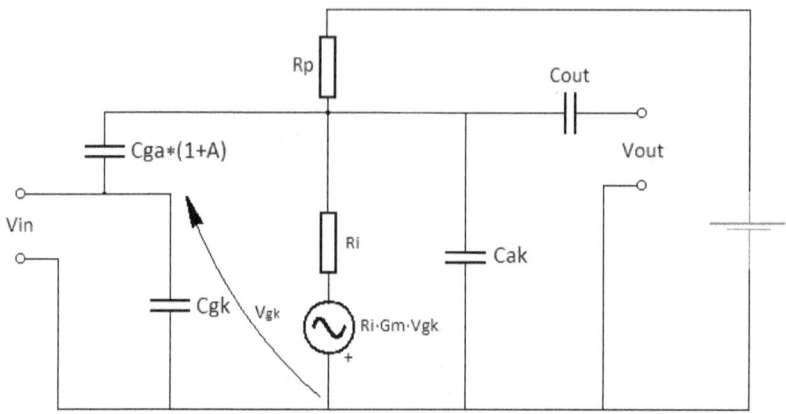

Fig. 24 – Equivalent model of a triode amplifier in class A with inter-electrode capacities.

The most important change is that the grid must be considered now as an electrode through which, alternating currents can flow. Hence, the grid is now part of two circuits in parallel; negative feedback loop and cathode leakage.

In view of the change in the effective capacity, according to the Miller effect, it is possible to state that:

"the bandwidth of the system is inversely proportional to the gain obtained".

Possible optimizations:

The inter-electrode capacities can not be eliminated or reduced except by changing the internal design of the electrodes (mainly by increasing its distance), but their impact on a circuit can be minimized in certain ways:

- The input capacity (grid - cathode) acts reducing the effective input impedance. In cases where offering a high value is required, an early stage with little (or unitary) gain acting as a "driver" can be added.

- The output capacity (anode - cathode) is in parallel with the load. If the load used has a much lower impedance than the reactance of this capacity throughout the working band, the effect can be neglected.

- The feedback capacity (grid - anode) produces a negative feedback loop that reduces the effective gain of the system as the working frequency grows. The only solution is to cascade several stages with lower individual gains. If the required bandwidth and gain are still not achieved, another tube model with lower capacities should be considered.

Operation of triode as amplifier

Class A inverting voltage amplifier

The most common use of a triode is for amplifying AC signals, i.e., accepting a variable input signal and delivering an output with similar characteristics and larger amplitude. The circuit most used for this purpose is the amplifier in Class A with anode output (also referred to as "common cathode"). The problems and limitations that arise when analyzing this topology apply to subsequent designs and provide a common basis for understanding more complex circuits such as complete amplifiers, phase inverters and voltage followers.

Basic idea and design criteria

Taking advantage of the relationship between the anode current and the grid-cathode voltage, a small level voltage can excite the tube varying a current that, through a load, will produce a larger amplitude voltage. At the same time, input and output impedances must be taken care of, usually making the former very large and the latter as small as possible. Finally, it must be checked if the amplifier can work properly throughout the desired frequency band, so internal capacities of the tube must also be taken in account.

Basic design criteria are then:

- system gain,

- high input impedance,

- low output impedance and

– working bandwidth.

The limitations of each criterion originate from different nature, namely:

System gain: the parameter limiting the most is the amplification factor of the tube itself. The circuit design can only maximize the inherent capabilities amplification of the model.

Input impedance: the impedance between grid and cathode is very large for mid frequencies when working with negative grid bias (grid voltage relative to the cathode), which is the usual case. However the grid-cathode capacity, may have a major impact for high frequencies. For example, for a typical 12AX7 triode, the grid-cathode capacity has a value close to 2 pF, so at 20kHz (typical audio band upper boundary), the reactance will have a value given by Equation 8.

$$X_C = \frac{1}{2 \cdot \pi \cdot 20000 \cdot 2 \cdot 10^{-12}} = 3.98 \, M\Omega \, [13], \qquad [8]$$

which, in practice limits its use for sources with very high output impedance (e.g. for a signal from a microphone capsule, but not from a microphone) but is suitable for other audio sources. One possible solution is to use an input transformer to raise the impedance seen by the signal source, with the reduction of amplitude level that entails (and possible additional problems caused by introducing a transformer).

Output impedance: mainly limited by the internal resistance of the tube and the maximum operating voltage, and which in turn limits the maximum anode load resistor that can be used. A good design

---

13 Reactance which is formed between grid and cathode due to the internal capacity. Results in a limitation on the maximum operating frequency of the tube.

can minimize the value of the output impedance, but the characteristics of the tube impose a fairly high minimum value by itself. The anode-cathode capacity of the tube introduces an additional limitation, since it acts as a lowpass filter, increasing the output impedance of the equivalent circuit for high frequencies.

Useful frequency band: the inter-electrode capacities limit the maximum operating frequency of the tube. They are formed between the electrodes as:

- grid - cathode: reducing the voltage existent between grid and cathode as the frequency of the excitation signal increases. Since the capacitive reactance takes small values, combined with the output impedance of the source it forms a voltage divider.

- grid - anode: produces a negative feedback (anode and grid are in phase opposition) that increases with frequency. When the frequency of the input signal reaches a high enough value, the feedback reduces the amplification to unity gain.

- anode - cathode: acts as a lowpass filter at the output, introducing an attenuation that increases with frequency and is independent of the load.

The operating limitations are strongly marked by the behavior of the tube, but the overall capabilities can be optimized with proper design. In the case where the performance obtained is still not sufficient to meet the design criteria, one should opt for another model with better specs. It is important to take into account the aging of components, because with the use, the amplification factor (transconductance) may decrease, so it is not a good idea to realize designs that require a tube to remain at its maximum performance. Such a design is not realistic for the normal working conditions and should be avoided whenever possible.

Circuit topology

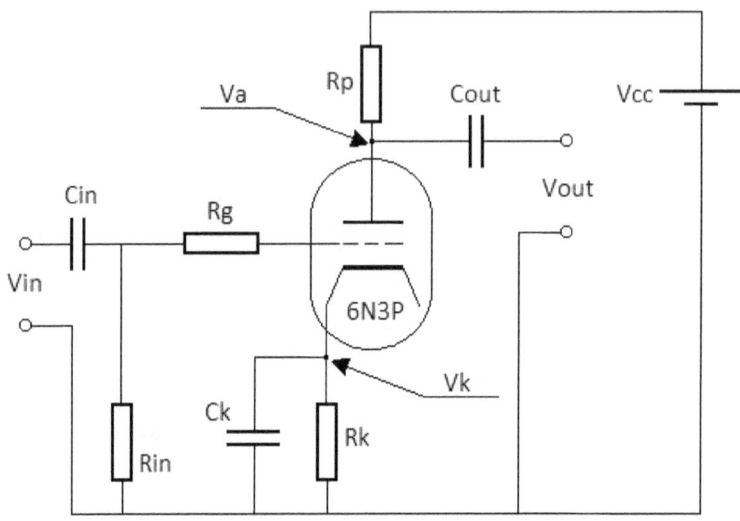

Fig. 25 – Schematic of inverting voltage amplifier using a triode with anode output and common cathode.

The circuit of Figure 25 shows the arrangement of a typical class A[14] amplifier with output from anode and common cathode. The input signal is applied to the grid decoupled by a capacitor to prevent that, the possible existence of DC component from the signal source modifies the operating point (bias) of the circuit. The output is obtained via a decoupling capacitor connected to the anode. This capacitor blocks the DC component, forming a high pass filter with the load.

---

14 It is important to remember that the class of an amplifier is not given only by the topology of the circuit. Also, some polarization conditions must be fulfilled.

Polarization (bias)

For the choice of polarization it is necessary to know the limiting working values of the tube and the maximum amplitude of the input signal. Assuming a typical case where the nominal input signal is line level:

$$+4 \text{ dBu} = 1.23 \text{ V}_{RMS} = 1.74 \text{ V}_{peak} \text{ [15]}$$

and adding a margin of 6 dB, considering the maximum input signal corresponding to:

$$+10 \text{ dBu} = 2.45 \text{ V}_{RMS} = 3.46 \text{ V}_{peak} \text{ [16]}$$

Moreover, as specified in the catalog, the maximum voltage between anode and cathode is 300 Volts, and the anodic peak current of 18 mA. The maximum power is 1.5 W however, considerably less than the product of maximum voltage times the maximum current, which would give a value of:

$$300 \text{ V} \cdot 18 \text{ mA} = 5.4 \text{ W [17]}$$

The explanation for these limits comes from the origin of each individual limits;

- The maximum voltage is determined by the separation

---

15 Rated line level input voltage of professional audio equipment. The suffix "u" of the decibel scale indicates that the reference is 0.775 Volts, from the voltage that appears on a load of 600 Ω when it dissipates one milliwatt.
16 Maximum input voltage for normal operation of the amplifier design. It is desirable to maintain a reasonable margin over the nominal level. The margin of 6 dB is a reasonable choice to minimize the chance of saturation.
17 Power dissipated in the tube as heat if its independent voltage and current limits were combined. Reaching this limit in transient conditions is feasible, but achieving it in stationary state would lead to the destruction of the tube by the inability to get rid of heat excess. Maximum operating limit values of the tubes are usually not as critical as for semiconductor components.

between anode and cathode and therefore the appearance of the electric arc between them, which could lead to the instantaneous destruction of the triode (the electrical impedance is considerably reduced if the arc is produced, allowing the circulation of a much higher current that electrodes are unable to withstand).

- The maximum current corresponds to the ability of the cathode to dissipate heat produced by Joule effect. In practical terms, a current higher than indicated limit will cause excessive heating of the cathode and a rapid deterioration of the same, reducing its ability to emit electrons and thus the triode performance.

- The maximum combined power is limited by the capacity of the entire unit to dissipate heat. Exceeding the limit stated by the manufacturer will cause the operating temperature to rise above nominal values. The consequences include drift of the characteristic curves, reduction of the service life and even its destruction by overheating of the electrodes. Using external cooling can compensate overheating, but this technique is generally limited to high power tubes in radio transmission equipment.

Knowing these values, the grid-cathode voltage is set at the value of the maximum peak voltage of the input signal, in this case 3.46 Volts ($\approx$3.5V). This will allow the maximum input signal not to drive the triode to positive polarization. At the same time, it is found that the maximum negative excursion will not drive the triode to the cutting area. For this, the family of curves provided by the manufacturer (Figure 26) is observed.

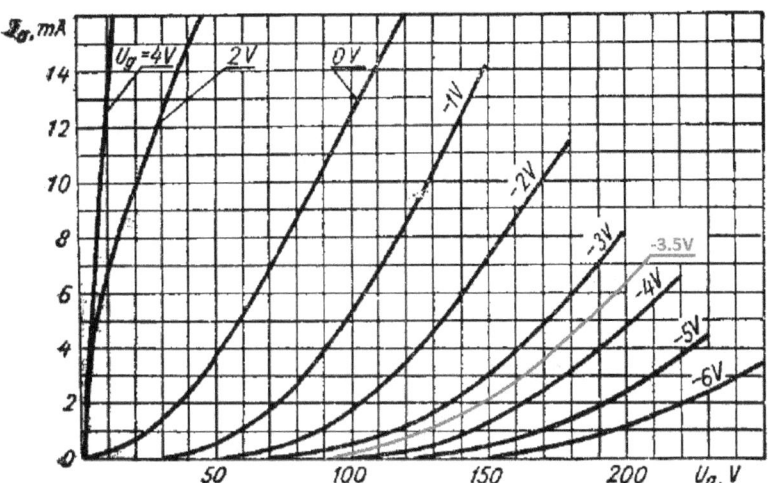

Fig. 26 – Family of characteristic curves of a triode model 6N3P. Source: datasheet from Russian manufacturer Reflektor. The curve corresponding to -3.5 Volts has been obtained as a mean of adjacent curves.

Maximum excursion will produce a negative peak in the grid of:

$$-3.46 \text{ (DC Bias)} + -3.46 \text{ (AC Peak)} = -6.92 \text{ Volts}$$

that is practically at the operating limit. Since the maximum positive peak will drive the grid voltage to 0 Volts, the operating range of the input signal is delimited as much as possible.

In conclusion, the bias voltage should be set at -3.5 Volts. This can be achieved in two ways:

– adding a DC component to the grid (for example, coupling it through a resistor), or

– inserting a resistor in the cathode thereof its voltage rises above the grid, so that the differential grid-cathode voltage is negative. This option is usually called "cathode-bias", "self-bias" or polarization by cathode resistor, and is the most common for circuits like this.

The value of the cathode resistor (R_k) must be such that, when no signal is applied to the grid, the grid-cathode voltage is -3.5 V. Then the DC component of the cathode voltage will always be +3.5 V. Starting from the -3.5 V corresponding curve and taking a central point of the most linear region, the current-voltage pair is 185 V and 5 mA, according to Figure 27.

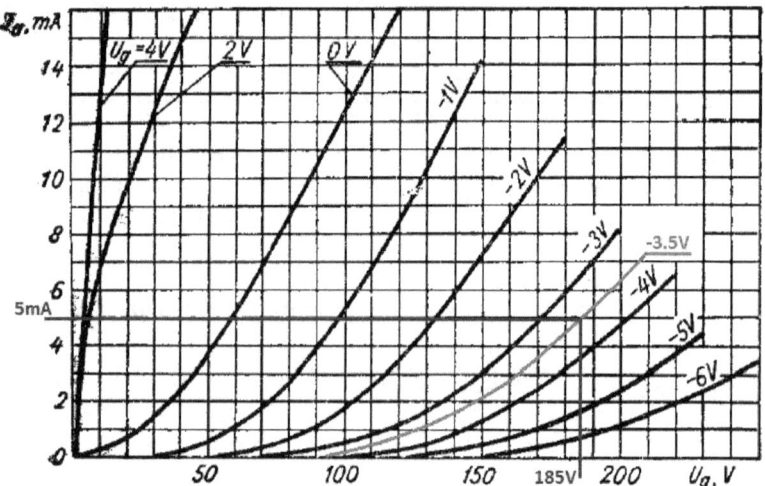

Fig. 27 – Bias point is selected (185 V / 5 mA) on the voltage polarization curve (-3.5 V grid-cathode).

Therefore, cathode resistor can be calculated as:

$$R_k = \frac{3.5V}{5mA} = 700\,\Omega \quad [18].$$

[9]

Once cathode resistor is obtained and operating point of the tube

---

18 The cathode resistor is chosen after the working point of the tube; polarization voltage of the grid (-3.5 V) and polarization current of the cathode (5 mA).

(185 V / 5 mA) is known, the next step is to calculate the resistance of the anode. It will produce the voltage fluctuation that will propagate to the output as the amplified signal through the bypass capacitor.

For this calculation it is necessary to apply the concept of internal transconductance and internal resistance. These may be calculated on the graph of the characteristic curves as:

- internal resistance: slope of the curve around the working point,

- transconductance: variation of anodic current with grid voltage for a given constant anode-cathode voltage.

The slope of the curve is obtained from two points and their respective coordinates (figure 28).

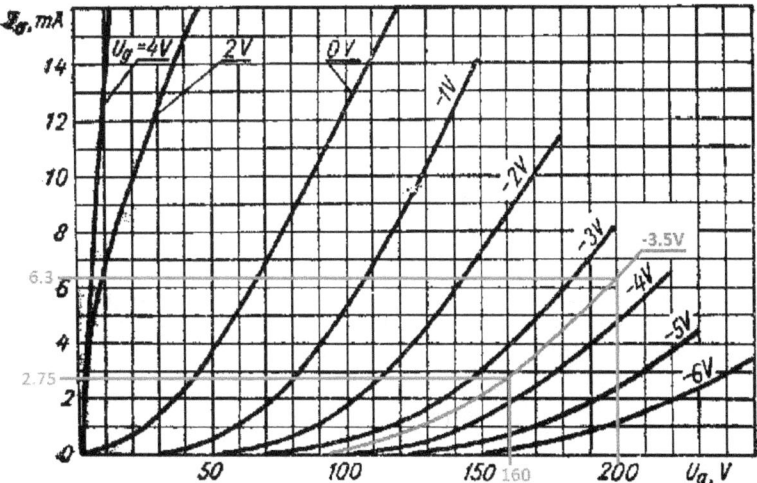

Fig. 28 – The internal resistance to the chosen working point is obtained as the slope of the curve around the selected point. If the variation of the output signal is very large, an appreciable error is introduced by considering internal resistance as a constant (rather than being a function of the anode voltage).

Such internal resistance follows Equation 10:

$$r_i = \frac{200V - 160V}{6.3mA - 2.75mA} = 11268\ \Omega \quad [19] \qquad [10]$$

The transconductance is calculated as in figure 29:

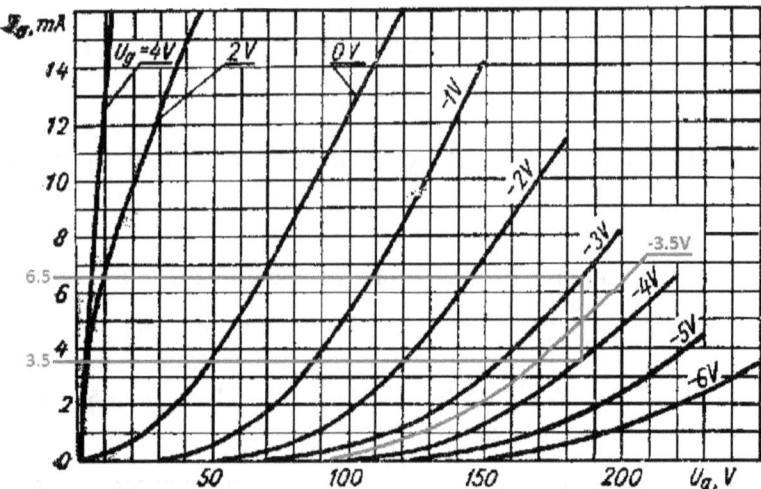

Fig. 29 – Transconductance is defined as the variation experienced by the anode current when the grid voltage varies one Volt and anode voltage is kept constant.

$$G_m = \frac{6.5mA - 3.5mA}{(-3V) - (-4V)} = 3mA/V \quad [20]. \qquad [11]$$

Knowing the values of these parameters, the amplification factor is obtained as (equation 12):

$$\mu = G_m \cdot r_i = 3mA/V \cdot 11268\ \Omega = 33.8\ V/V \quad [21]. \qquad [12]$$

---

19 The internal resistance is a model abstraction, useful for calculation but does not physically exist in the tube.

20 Calculating the transconductance should be made with reduced operating margin to avoid introducing excessive error.

21 The amplification factor is the only model parameter which remains relatively constant, while the transconductance and internal resistance are highly dependent on the operating point.

Based on the known polarization conditions, where the anode voltage is 185 Volts, the maximum signal excursion will cause anode voltages of:

$$185V\,(DC)+(33.8\,V/V\cdot3.5V)=303.3\,V \quad ^{22}\text{(positive}$$
$$\text{output peak)}$$

y

$$185V\,(DC)+(33.8\,V/V\times-3.5V)=66.7\,V \quad ^{23}\text{(negative output}$$
$$\text{peak).}$$

These results fit almost perfectly to the maximum value allowed by the tube design, which has a limit to the anode voltage of 300 Volts. Therefore, the high voltage supply will be the maximum allowed: 300 Volts, and the anode resistance that   meets the condition polarization conditions (equation 13).

$$R_a=\frac{300V-185V}{5mA}=23\,k\Omega \quad ^{24}. \qquad [13]$$

At this point, it only remains to check that the limits of tube operation are not exceeded under static conditions (bias). The power dissipated when no input signal is present will be:

$$P_a=V_a\times I_a=(185V-3.5V)\times5mA=0.91W \quad ^{25}, \qquad [14]$$

a value within the range specified by the manufacturer; 1.5 Watts.

---

22 The maximum and minimum peak output are assumed following the linear model.
23 In the vicinity of the working limit, due to the nonlinearities of the tube, smaller excursions will be obtained. Thus considerable deformations are introduced to the signal.
24 The anode resistor affects the frequency response in addition to the polarization of the circuit, as will be seen later.
25 It is essential to check that the tube does not exceed its maximum power on stationary conditions.

Typically, the power limit is briefly exceeded on transient conditions, and this should not pose danger to the tube. However, it is not acceptable a design where as stationary condition the maximum power is dissipated.

Input circuit

The purpose of the coupling circuit at the amplifier input is simply to eliminate the possible DC component of the original signal and provide a controlled input impedance.

The capacitor C_{in} prevents DC components from driving the grid, and its value should be such that the high-pass circuit that forms along with R_{in} filter does not interfere with the working band (20 Hz – 20 kHz considering the audio band). Locating the cutoff frequency below the lower limit of the band (20Hz) is adequate, but it must not be forgotten that the phase of an RC filter as mentioned begins to change a decade before and stabilizes a decade after that cut frequency. If a work bandwidth with a linear phase is required, the cutoff frequency must be at least one decade below 20 Hz; i.e. 2 Hz.

The value of R_{in} is dictated entirely by the required amplifier input impedance for a center frequency (1 kHz). As an amplifier that will accept a line signal, the default is 10 kOhm.

The value of capacitor C_{in} is obtained from the expression of the cutoff frequency of a first order RC high pass filter:

$$f_c = \frac{1}{2 \cdot \pi \cdot R \cdot C} = \frac{1}{2 \cdot \pi \cdot 10k\Omega \cdot C} = 2\,Hz \quad [26], \qquad [15.a]$$

---

26 Do not forget that, at the cutoff frequency, the module has reached -3 dB respect to the passband, which represents a ~30% lower, with a soft "elbow"

then: $C=\dfrac{1}{2\cdot\pi\cdot 10k\,\Omega\cdot 2Hz}=7.96\mu F\approx 8\mu F$ [27]. [15.b]

The choice of input capacitor is critical, and it should be of the highest possible quality, for any introduced defects will have a major impact in the overall performance. Large capacities are often required, making it more difficult to find them on high quality capacitor types. If a higher value of R_i can be used (e.g. 100 kΩ), the required capacity will drop to (equation 15.c):

$$C=\dfrac{1}{2\cdot\pi\cdot 100k\,\Omega\cdot 2Hz}=796nF\approx 800nF \quad [28], \quad [15.c]$$

a value that can easily be found in polyester, polypropylene or metal film capacitors, more desirable for audio.

With these two values, the input section is configured, except for the resistance of R_g whose purpose is to prevent the so called "grid blocking". In the event of the input signal driving the grid-cathode voltage to positive values, electrons traveling from the cathode to the anode may adhere to the grid. The current produced can severely overdrive the tube introducing considerable distortion. By placing the resistor R_g, this current is reduced. Under normal working conditions, the polarization will be negative and the current flow trough the grid effectively zero, so the presence of R_g will introduce no actual effect.

---

approaching, being a first order RC filter.

27 The use of large capacitors in audio applications can be a problem primarily by the nonlinearities in phase these devices may introduce.

28 Using components of close but not exact values is common. The circuit design must ensure the normal operation of the system even with small values differences (plus the natural aging components will suffer).

Output circuit

The variation of the anode current following changes of the input voltage present on the grid, will produce the voltage drop at the anode resistance to be also variable and inversely proportional to the input voltage: a rising input voltage will produce an increase in the current conducted by the tube and the resistance on the anode, and therefore the voltage drop will be higher. This is how the phase inversion occurs in the amplification generated by this circuit.

The capacitor C_{out} forms a first order high pass RC filter with the load. The value of this is not known, but in many cases it will be the input of another amplifier stage. In that case, it is appropriate to assume case an input impedance of 300 kΩ or higher for midrange frequencies (1 kHz). Accordingly, the value of C_{out} should be such that the cutoff frequency is below 2 Hz:

$$C_{out} = \frac{1}{2 \cdot \pi \cdot 2\text{Hz} \cdot 300\text{k}\,\Omega} = 265\text{nF} \quad [29]. \qquad [16]$$

This value must be adjusted to the load when known, and if not, the highest practical value.

Cathode capacitor C_k does not exactly belong to the output circuit, but its presence affects the output, so it is necessary to briefly discuss its situation in this section. Its mission is to maintain polarization constant despite AC currents which occur as a result of the alternating signals in the grid. Should it not be present, dynamic and

---

29 When load is of unknown value, the output capacity should be maximized to avoid losses at low frequencies.

harmonic distortion would be introduced. By placing it, the cathode remains at a fixed DC voltage, allowing a much more stable working point. There is not a unique criterion to select the capacitor value, but a good approach is to consider the circuit a half wave rectifier filter. Being the DC current that corresponding to the bias, the peak to peak voltage the maximum input swing and the output ripple reduced by a factor of ten. If so, capacity should be at least of:

$$C_k = \frac{I_{DC}}{f \cdot \dfrac{V_{PP}}{10}} = \frac{5\text{mA}}{20\text{Hz} \cdot \dfrac{7\text{V}}{10}} = 178 \mu F \quad [30], \qquad [17]$$

being the maximum swing of the input signal:

$$V_{PP} = 2 \cdot (V_{KG(DC)}) = 2 \cdot (3.5) = 7\text{V} \quad [31].$$

---

30 Cathode capacity is not a too critical value. It is interesting to obtain frequency response data (depending on the cathode capacity) in each circuit under development when possible. In some literature, cathode capacity is expressed as the time constant of the $R_k - C_k$ circuit. Both expressions are equivalent. It is also interesting to note that the ESR of the C_k may have an important effect on the transient response of the amplifier.
31 Input signal will make the grid-cathode voltage to swing (alternating component) around the bias point (continuous).

Final calculations

Gain of the amplifier

From the amplification factor, previously known, the gain of the complete amplifier can be obtained. The effect of the input and output circuits must now be included.

The input filter formed by C_{in} and R_{in} introduce an attenuation described by the voltage divider:

$$Att_{in} = \frac{V_{out}}{V_{in}} = \frac{R_{in}}{R_{in} + X_{Cin}} \quad {}^{32}, \qquad [18.a]$$

considering mid frequencies (1 kHz):

$$Att_{in} = \frac{10k\,\Omega}{10k\,\Omega + \dfrac{1}{2 \cdot \pi \cdot 1kHz \cdot 8\mu F}} = 0.99 \quad {}^{33}. \quad [18.b]$$

It can be said that the input circuit is transparent for the calculation of the gain for mid frequencies. The loss introduced can therefore be neglected.

The output filter formed by C_{out} and the load and responds to the same expression, and if the value of the load is the expected (100 kΩ or higher), the introduced attenuation is negligible.

What remains is to evaluate the gain of the tube circuit to determine the gain of the complete amplifier. The equivalent model

---

32 Attenuation (times) of the input filter. The passive coupling circuits of input and output introduce an attenuation that is necessary to evaluate.
33 The attenuation depends on the frequency. The filter is transparent for all passband frequencies.

conveniently serves this purpose. The circuit, at medium frequencies (considering only alternating signals) is as follows:

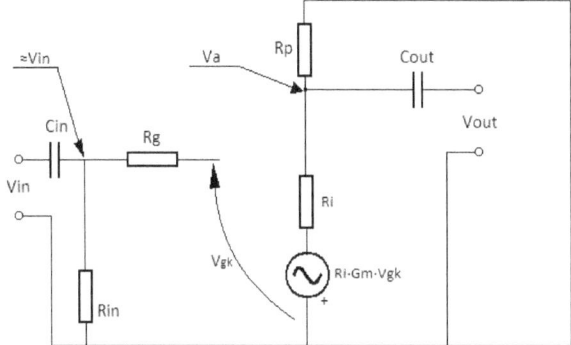

Fig. 30 – Equivalent circuit of amplifier for mid frequencies considering only AC components.

During the AC analysis, the high voltage supply becomes shorted to ground because it is a DC voltage source.

At the same time, the cathode condenser makes the effect of bypassing the cathode resistor, so this pair may be replaced by a short circuit to ground. The circuit is substantially simplified (Figure 31).

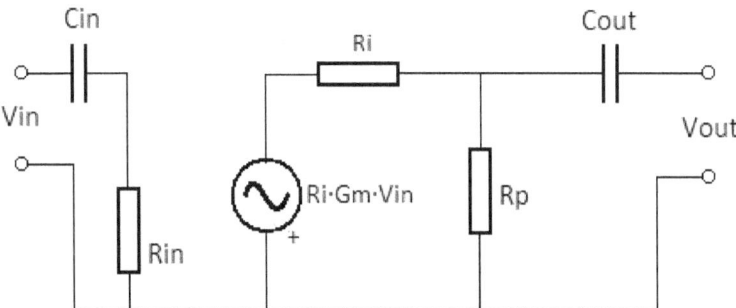

Fig. 31 – The amplifier in a simplified form, as a voltage generator depending on the input signal and a resistive voltage divider. This simplification is valid for mid frequencies.

And substituting the known values of the equivalent model:

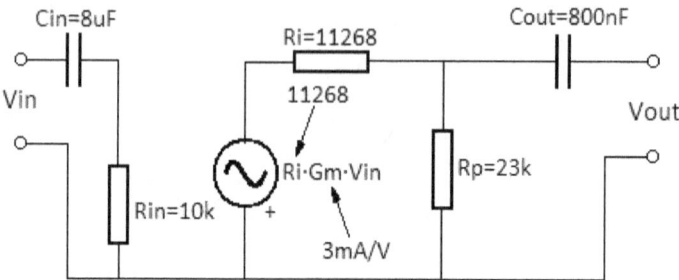

Fig. 32 – After substituting the values of the equivalent model obtained from the curves, the resolution is simple.

The ratio input - output is now clear and can be obtained from the voltage divider formed by R_i and R_p (disregarding the effect of C_{out}, because the load is unknown) as equation 19.a.

$$V_{out} = R_i \cdot G_m \cdot V_{in} \frac{R_p}{R_p + R_i} \qquad [19.a]$$

$$V_{out} = 11268\,\Omega \cdot 3\text{mA}/V \cdot V_{in} \frac{23\text{k}\Omega}{23\text{k}\Omega + 11268\,\Omega} = 22.68 \cdot V_{in}$$

a gain that can be expressed in decibels (Equation 19.b).

$$Gain\,[dB] = 20 \cdot \log\left(\frac{V_{out}}{V_{in}}\right) = 20 \cdot \log(22.68) = 27.1\, dB \quad [34]$$

$$[19.b]$$

This value applies to mid-frequency signals. At low frequencies (below 2 Hz) the input and output filters will introduce significant attenuation and the resulting gain will be lower. The same occurs at high frequencies, where the inter-electrode capacity of the tube will act as a low pass filter limiting the effective gain that can be obtained from the amplifier.

---

34 The gain in decibels provides a better idea of the efficiency of the amplifier.

Frequency response – Operating bandwidth

At low frequency, the limit is set by two aspects of the circuit, the input filter formed by C_{in} and R_{in} and the output filter formed by C_{out} and the load. Both have been previously studied to achieve an adequate response from 20 Hz on. At high frequency, the inter-electrode capacities introduce losses in the gain that can be obtained from the amplifier (mainly due to the Miller effect). The capacities that occur are:

- Capacity grid - cathode (C_{gk}) forms a low pass filter on the input, reducing the effective voltage that the triode receives,

- Capacity grid - anode (C_{ga}) creates a negative feedback loop, so that the higher the operating frequency, the more output signal is injected back at the input. Because of that, above a certain frequency the amplification will be null. The effective value of this capacity grows as many times as the circuit gain. According to Miller with a factor of $(1 + A)$ times, where A is the gain.

- Capacity anode - cathode (C_{ak}) forms a low pass filter at the output, reducing the amount of signal that is delivered to the load.

The arrangement of capacities at the equivalent circuit are:

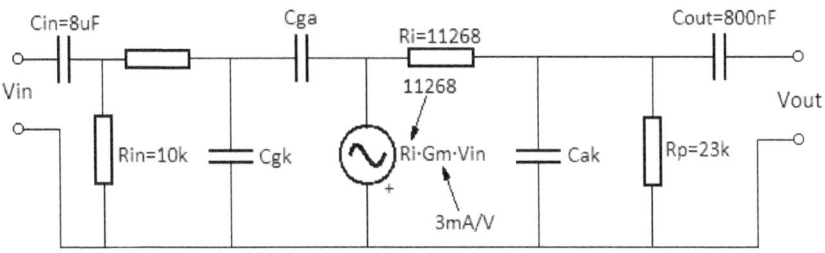

Fig. 33 – The equivalent circuit must consider the internal capacities of the tube. These limit the system operation at high frequencies. The main limit is imposed by the grid - anode capacity for generating a negative feedback loop and being effectively increased due to the Miller effect.

The capacity values are provided by the manufacturer, and according to the data sheet, each triode encapsulated in a 6N3P has:

C_{ga} = 1.6 pF→ effective value due to Miller: C_{ga}* = **1.6 pF · 22.68 = 36.3 pF**

C_{gk} = 2.8 pF

C_{ak} = 1.4 pF

To better understand the effect of inter-electrode capacities and their effect on high frequency response, the circuit can be simplified to Figure 34.

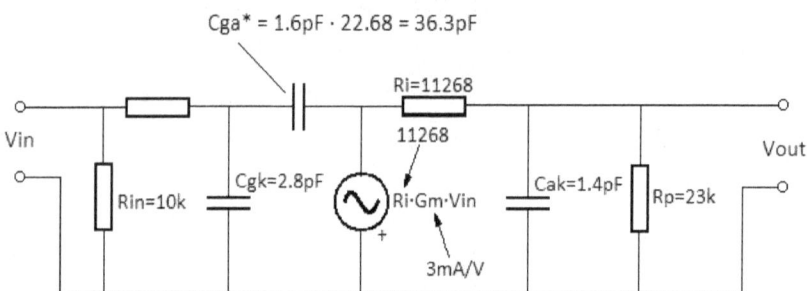

Fig. 34 – At high frequencies, the input and output filters can be simplified as a short circuit for ease of the analytical study of the circuit.

The main limitation is always imposed by the grid - anode capacity since it produces a feedback loop that causes an unsurpassable attenuation that increases with frequency.

The effect of the input capacitance (C_{gk}) can be minimized by using a low output impedance signal source (or inserting a stage that provides the necessary impedance adaptation).

The attenuation of the output capacity (C_{ak}) can not be minimized, but a load of low value can be used. Being in parallel with the X_{Cak} reactance, it will maximize the operating band at high frequencies (in exchange for a lower gain in the whole band).

Simulation of equivalent circuit

The complete amplifier circuit, treated as the equivalent model is suitable for analysis by computer simulation software. It will provide a full frequency response with magnitude and phase.

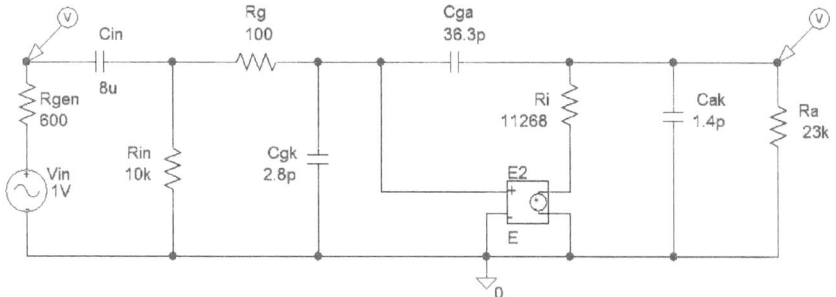

Fig. 35 – Computer circuit simulation. The use of specialized software simplifies the design work avoiding performing tedious iterative calculations.

Using any electronics simulation software, the gain of the system (output voltage as a function of the input voltage) can be obtained (Figure 36).

The previous calculations are confirmed;

- The passband covers the central area of typical audio spectrum (20 Hz – 20 kHz) with a gain of 27.1 dB.

- The cut in the low frequency is approximately at 2 Hz (set by the input filter formed by C_{in} - R_{in}).

71

–   At high frequencies, inter-electrode capacities limit the gain obtainable by the tube in the circuit. The cutoff frequency is located in approximately 250 kHz. The phase will begin to change from 25 kHz on.

Fig. 36 – Module of the frequency response in dB. The passband comprises the entire audio spectrum. Cutoff frequencies are defined by the input filter (low frequency) and the inter-electrode capacity (high frequency).

For audio applications, besides a flat enough module in the frequency response, it is important to maintain a linear phase response. At low frequencies the cutoff frequency has been carefully set to a decade before the start of the band, to ensure that the phase is stable before reaching the operating band (starting at about 20 Hz). In high-frequency, the limitation is imposed by the internal capabilities of the tube, so the result can only be observed and if they are unsatisfactory, use another model with better internal electrostatic insulation. An analysis of the relative input - output phase is shown on Figure 37.

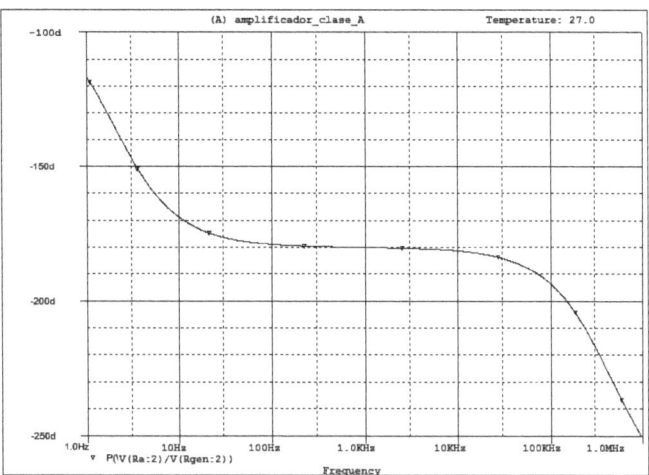

Fig. 37 – Phase of the frequency response of the complete amplifier. Phase changes cover two decades around the cutoff frequencies. That is why if a response is required to be without any deformation, cutoff frequencies must be at least a decade away from the boundary of the passband.

From the plot of the phase of the frequency response, the following conclusions are extracted:

–   The passband has a constant phase difference of 180 degrees with respect to the input (it is inverted).

–   At the low frequency limit (20 Hz), phase has changed linearly to a value of +5 degrees relative to the central area. This deviation is acceptable.

–   At the high frequency limit (20 kHz), phase has changed linearly to a value of -5 degrees relative to the central area. Again an acceptable value.

The overall result of the frequency response is positive, and has demonstrated the influence of the different parts of the circuit thereon; the input and output filters, and internal capacities of the tube.

73

From the equivalent circuit with inter-electrode capacities, the one with the most weight is that formed between grid and anode, for producing a negative feedback loop whose impedance decreases with the frequency (more negative feedback at high frequencies). To test this assumption, the C_{ga} capacity has been eliminated from the circuit (has been replaced by an open circuit) and the simulation of frequency response repeated (Figure 38).

The result is obvious; high frequency cutoff moves above 1MHz (the limit of the simulation performed). Now grid-cathode and anode-cathode capacities are ones limiting the response at the upper end of the spectrum.

Using a tetrode or pentode in its place, the amplifier would have been able to extend the high frequency response with ease, as these devices have much lower grid - anode (C_{ga}) capacities.

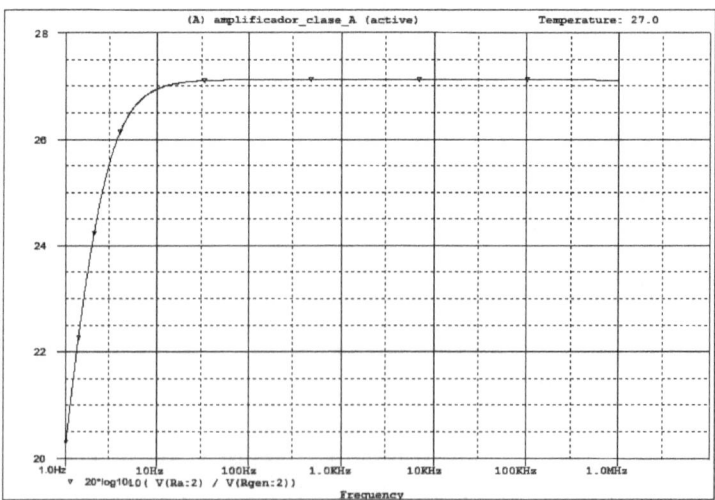

Fig. 38 – Module of frequency response in the case where the grid - anode capacity was null. It is found that this is the greatest limitation introduced by the inter-electrode capacity.

It should not be forgotten that this model of tube (6N3P) is very well suited for audio, where a response as obtained (Figure 36) is considered in High Fidelity equipment. For cases where it is required to extend the operation at higher frequencies, there exist other triodes designed for this purpose.

Estimated harmonic distortion

The lack of linearity in the tube translates to the introduction of distortion (mainly described as: harmonic and intermodulation). Because the simulation is performed with a linear model the amount of distortion introduced can not be estimated thereby. There are, however, expressions that can estimate the distortion introduced by a triode functioning as a class A amplifier.

The triode produces mainly second order harmonic distortion, so the THD (total harmonic distortion) depends largely on the second component. The expression for calculation is:

$$HD(2°) = 100 \cdot \frac{\Delta V_{a2} - \Delta V_{al}}{2 \cdot (\Delta V_{a2} + \Delta V_{al})} \ ^{35}, \qquad [20.a]$$

being ΔV_{a2} and Δv_{al} the differential anode voltages reached when the grid voltage is varied in +1 and -1 Volts respect to polarization value. According to the characteristic curves, anode voltage will vary following figure 39.

If, on the load line (diagonal, descending line), the grid voltage is shifted to a one less volt (-4.5 V), the anode voltage will increase to 208 V, and if it does to a one more volt (-2.5 V), the anode voltage will drop to 160 V.

---

35 Formula for estimating the second order harmonic distortion introduced by the triode.

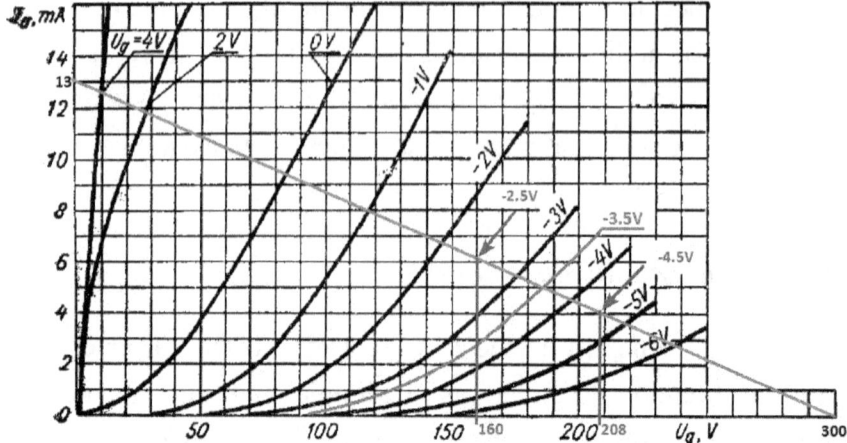

Fig. 39 – Family of characteristic curves with the corresponding load line (diagonal, descending line). The variation of the anode voltage is related to the variation of the grid in a quasi – linear fashion.

The expression for estimating the second order harmonic distortion results in Equation 20.b:

$$HD\left(2^{\circ}\right)=100\cdot\frac{\Delta V_{a2}-\Delta V_{a1}}{2\cdot\left(\Delta V_{a2}+\Delta V_{a1}\right)}=100\cdot\frac{25-23}{2\cdot\left(25+23\right)}=2.08\%\quad [36]$$

[20.b]

This value is estimated, and should be interpreted as a reference. The only way to obtain accurate data of the harmonic distortion is by taking actual measurements on the circuit.

---

[36] Estimation of second order harmonic distortion introduced by the non linear character of the curves.

Note: explanation of the load line

Knowing that the load of the circuit in AC is due only to the series resistor R_a, it is possible to draw the load line by finding two points:

- Assuming that the tube is on cutoff (anode current is null), output voltage will equal to that of the supply, given that there will be no voltage drop on R_a. Hence, one of the points will be 300 V / 0 mA.

- Assuming that the tube is completely saturated (anode-cathode voltage is zero), the current through the load R_a will be, according to Ohm's law:

$$I_{Ra} = \frac{300\text{V}}{23\text{k}\,\Omega} = 13\text{mA}\ ^{37},$$

hence, the second point will be 0 V / 13 mA. The load line cuts the chosen curve (-3.5 V) at the known polarization point: 185 V / 5 mA.

Using the load line it is possible to know, by simple observation, the variation the anode voltage will experience as a function of the grid voltage.

---

37 Calculation of load line of the circuit.

Bibliography

GRAY, T. S., Applied Electronics, 9th edition, MIT Press, Cambridge, 1965.

BRIGGS, G. A., Amplifiers, the why and how of good amplification, 1st edition, Wharfedale Wireless Works, Bradford, 1952.

G.E.C VALVE AND ELECTRONICS DEPARTMENT, An approach to audio frequency amplifier design, 1st edition, Chapman & Hall Limited, London, 1957.

YORK, H. L., Amplifiers, the technique of sound reproduction, 1st edition, Focal Press, London, 1964.

SCOTT, R. F. and others, Amplifier builder's guide, 3rd edition, Radio Craft, New York, 1947.

WILLIAMSON, D. T. N., The Williamson amplifier, 2nd edition, Wireless World, London, 1952.

DICKIE D. P. y MACOVSKI A., A transformerless 25 watt amplifier for conventional loudspeakers, 1st edition, unknown, 1954.

BAXANDALL P. J., Negative-feedback control, 1st edition, Wireless World October 1952.

HAMBLEY, A. R., Electronica, 2nd edition, Pearson, Madrid, 2008.

VARIOS, Introducción a la electrónica, 1st edition, Ceac, Barcelona, 1977.

VARIOS, Improvisaciones que dan dinero y ahorran tiempo, 1st edition, Ed. Técnicas Rede, Barcelona, 1966.

Also, several tube datasheets have been used, from the following brands: Tung-Sol, Reflektor, General Electric, Silvania, Svetlana, Phillips, Marconi-Osram y JJ.